AF832718

THE SOUTHERN BOUNDARY REGION OF THE WINTER ANOMALY IN IONOSPHERIC ABSORPTION IN WINTER 1971/72 OBSERVED ON BOARD THE CARGO VESSEL "HANAU" OF HAPAG-LLOYD MOVING BETWEEN 10° AND 55°N

I. Measurements and Results

by

W. ELLING, K.H. GEISWEID, and H. SCHWENTEK

II. An equipment for an A1 mobile station

by

W. BARKE, K.H. GEISWEID, G. HEIMESAAT,
A. LOIDL, C. RÖMER, and R. ZELLERMANN

Diese Mitteilungen setzen eine von Erich Regener begründete Reihe fort, deren Hefte am Ende dieser Arbeit genannt sind.

Bis Heft 19 wurden die Mitteilungen herausgegeben von J. Bartels und W. Dieminger. Von Heft 20 an zeichnen W. Dieminger, A. Ehmert und G. Pfotzer als Herausgeber, ab Heft 51 W. Dieminger und G. Pfotzer.

Das Max-Planck-Institut für Aeronomie vereinigt zwei Institute, das Institut für Stratosphärenphysik und das Institut für Ionosphärenphysik.

Ein **(S)** oder **(I)** beim Titel deutet an, aus welchem Institut die Arbeit stammt.

Anschrift der beiden Institute:

3411 Lindau

ISBN-13: 978-3-540-06660-6 e-ISBN-13: 978-3-642-65820-4
DOI: 10.1007/978-3-642-65820-4

Contents

I. Measurements and Results

1. Introduction

In the framework of the European Cooperation concerning the study of the geographical distribution of the winter anomaly in ionospheric absorption besides the network of fixed stations covering the area from Spain (4°W) to Japan (141°E), at latitudes between 36° and 54°N, a measurement was planned and carried out to extend the geographical coverage of the study from Europe to the Gulf Coast of the USA by using a ship. LAUTER and SCHÄNING [1970] have shown that by choosing of a suitable route also the southern boundary of the occurrence of the winter anomaly could thus be determined. We supposed that this boundary is not fixed but changing from winter to winter, perhaps from month to month, or even from day to day; moreover, it cannot be excluded that it depends on geographical longitude.

Basing on the experience obtained from the measurement of absorption on the Norddeich-Lindau circuit for many years [SCHWENTEK, 1958, 1966, 1971], the optimal period for such an investigation would be the months December and January. A retrospective survey on the behaviour of absorption in Winter 1971/72 excellently verified our suggestion (Fig. 1). We started the expedition at the end of November 1971. Due to strike at the US harbours we had to use another route than previously had been planned, and went to Venezuela using a regular cargo vessel.

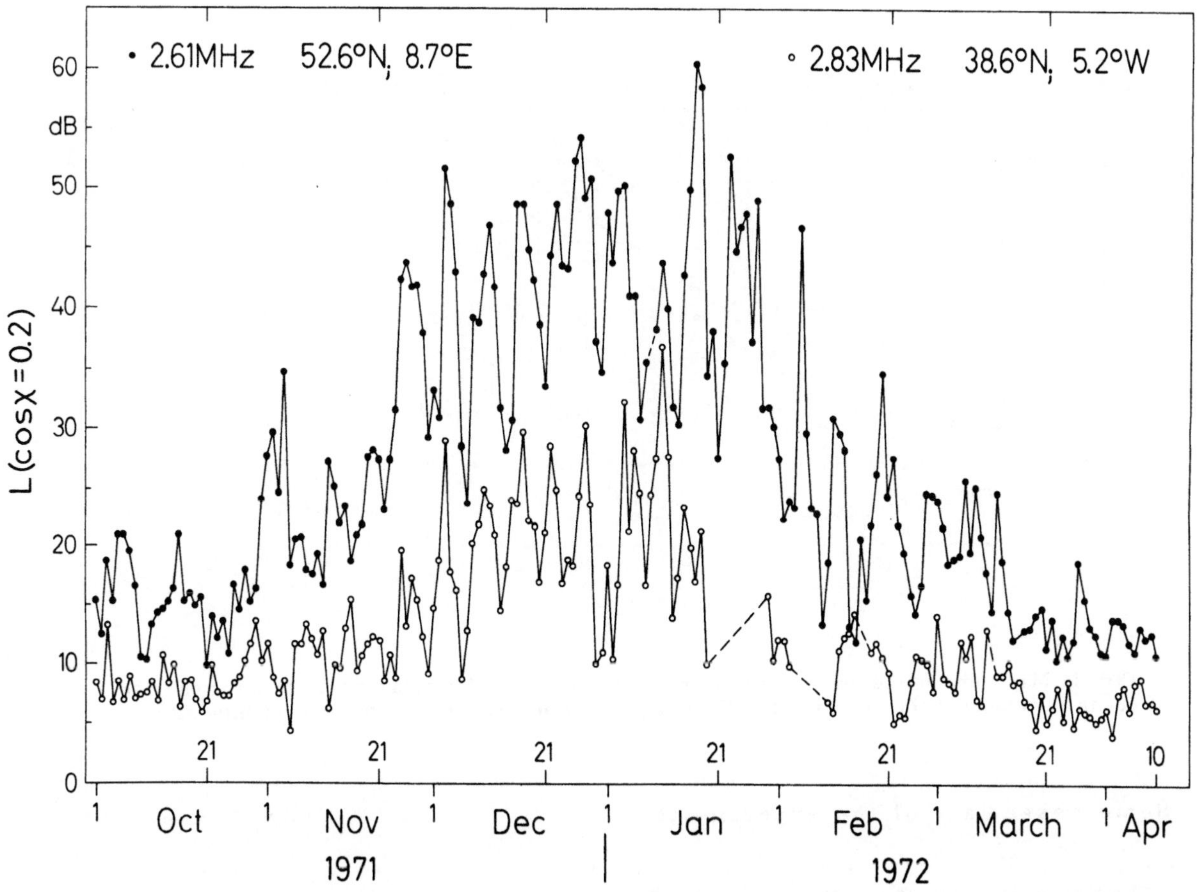

Fig. 1: Trend of ionospheric absorption L(cos χ = 0.2) in winter 1971/72. Circuits Norddeich-Lindau (2.61 MHz) and Aranjuez-Balerma (2.83 MHz). Note that the winter anomaly at the southern station is less pronounced, but similar; moreover, that in the trend of the absorption data a clear wave-like structure appears.

Starting at Hamburg (53°N) on 28 November 1971 measurements were made almost continuously, except in harbours where the antenna had to be put down. This was a certain disadvantage because in the European area the A1 data obtained on board should be used for comparison with the data of other stations, particularly, the A3 circuit Norddeich-Lindau. Nevertheless, some $L(\cos\chi = 0.2)$ values, (χ = sun's zenith angle), were obtained from those parts of the diurnal variation of absorption which could be measured when moving between harbours.

Besides the network of stations measuring absorption there were three launching sites for rockets at which also investigations of the lower ionosphere were carried out: South Uist (United Kingdom), Arenosillo (Spain), Wallops Island (USA). A measurement of absorption on board the ship near South Uist during the time of rocket launchings could not be realized because the time schedule of the ship was fixed and could not be changed.

Finally, a D region program was carried out at Wallops Island (USA) during the end of January and beginning of February. Therefore, we started again on January 23, going back to La Guaira (10°N), crossing the North Atlantic a third time (Fig. 2). The end of measurements was on 12 February 1972 at La Guaira.

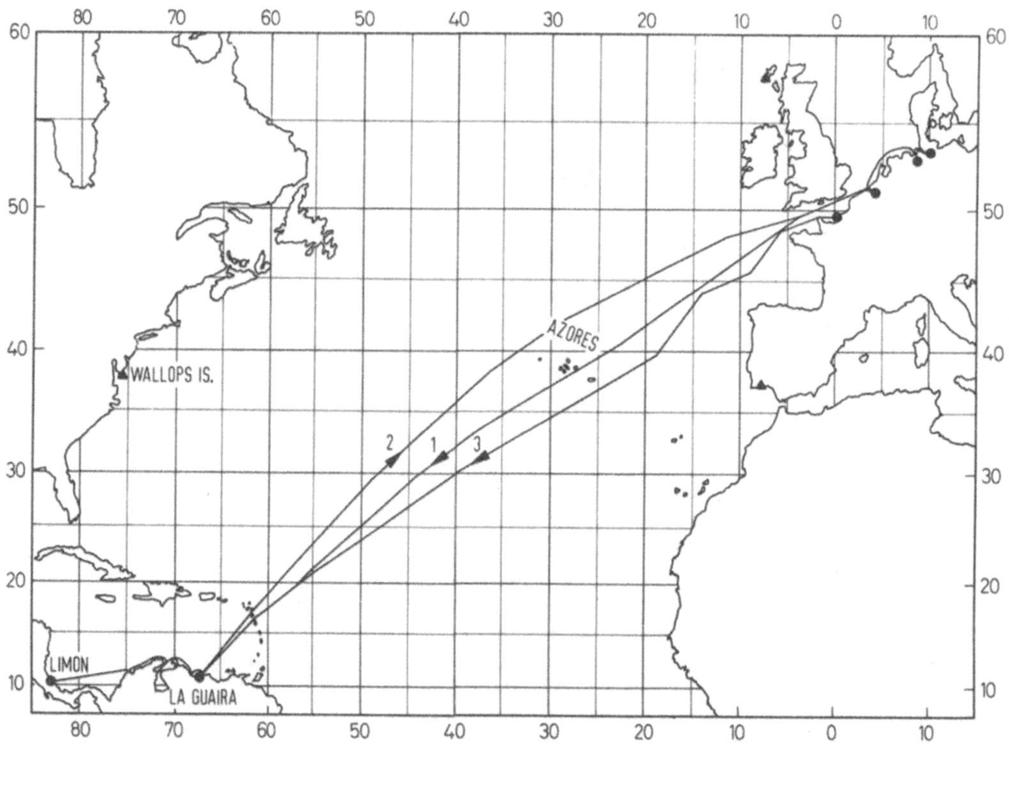

Fig. 2: Map showing the three routes of mv "Hanau" on which measurements were carried out in December 1971, January and February 1972; • harbours, ▲ rocket launching sites.

2. Some essentials of the experiment

The first application of our new mobile equipment was on a cargo ship which implied some limitations concerning the antenna and radiated power. In order to have a frequency as closely as possible to the mean equivalent frequency of the Norddeich-Lindau circuit (1.5 MHz for the 1E mode), we used one of the standard frequencies for absorption measurements at vertical incidence, 1.71 MHz; moreover, 5.06 MHz, a frequency just three times greater than 1.7 MHz, thus allowing without difficulties to use one folded dipole for the two frequencies. During daylight, waves of 5 MHz are normally reflected at the F layer, (and penetrating the E layer, if there is no strong Es), while waves of 1.73 MHz are reflected at the E layer. Thus, using also 5 MHz, the influence of the E layer on absorption could be checked. A difficulty, however, was the determination of a reliable nighttime value at 5 MHz for calibration.

The equipment used for the absorption measurements on board m.v. "Hanau" is described in some detail in part II. The concept was to develop a pulse transmitter with a peak power of 60 kilowatts for a simultaneous measurement of up to 10 fixed frequencies in the range from 250 kHz to 6 MHz, a receiver covering linearly 60 dB, a programmer allowing for many different programs, and a simple but reliable recording system providing data in such a way that they would be available on the expedition shortly after the measurement, that is, with no more delay than 24 hours.

The pulse length used was 200 μs, due to some troubles, the transmitter's pulse power much less than 60 kilowatts. The received pulses were displayed on an A-scope and photographed superimposing always 60 pulses within 2 min in the daytime, or 10 minutes during the night. A detailed description of this photographic pulse-superimposing method is given in the next section.

By courtesy of the shipping company a special mast could be mounted to lengthen one of the ship's masts by 8 m as a support for the dipole (Fig. 14).

3. Determination of the night-time calibration value, and description of the photographic pulse-superimposing method

Outlining the determination of the night-time value, simultaneously the photographic pulse-superimposing method will be described. - Since absorption L is defined by

$$L(f) = 20 \log \frac{h'_n \cdot E_n}{h'_d \cdot E_d} \quad \text{in dB}, \qquad (1)$$

where f frequency, h'_n and h'_d virtual heights of reflection at night or in the daytime, respectively, E_n and E_d amplitudes of the 1E or 1F reflection at night or in the daytime, respectively, the accuracy of the absorption data L depends strongly on that of the night-time value $h'_n\,E_n$. In our case, the reliability of the measured E_n value depended not only on the degree of the constancy of the equipment but might have been depended also on weather at sea by influencing the matching of the aerial when there was much spray in the air, and also on the position of the derricks which was not the same during the three passages. Therefore, for sake of reliability, two persons analysed independently the night-time records, during the first journey already on the ship, applying different methods.

Between sunset and sunrise, due to low critical frequency of the F layer, at 5 MHz nearly no signal was received; then, at 1.7 MHz only, every ten minutes, a picture with 60 superimposed pulses (24 mm x 36 mm) was taken on film (pulse-repetition frequency of 0.1 Hz). Next day, by means of a normal slide projector mounted suitably to get the picture on a table, one person determined the single amplitude and height values and listed median amplitudes E_n and median heights h'_n, the other put the values directly from the film into a desk calculator with plotter plotting the variation of the night-time values with time, and drawing a linear regression line for each night.

The listed data were analysed with regard to different conditions in the ionosphere, that is, separated for F reflections at a rather constant and reliable layer, and multiple reflections from the Es layer.

Then, median values were evaluated for the different samples of data. Thus, we tried to achieve the highest possible accuracy of the night-time value, and an estimate of its possible variation which is less than 3 dB.

Some typical records are presented in Fig. 3 to demonstrate the photographic pulse-superimposing method, and to give an impression of its usefulness. The height of reflection can be determined from the records with an accuracy of about ± 1 km. A check of the height scale was possible when, in the night-time, there was a blanketing Es layer causing multiple reflections, 1 Es, 2 Es, 3 Es (see Fig. 3.2).

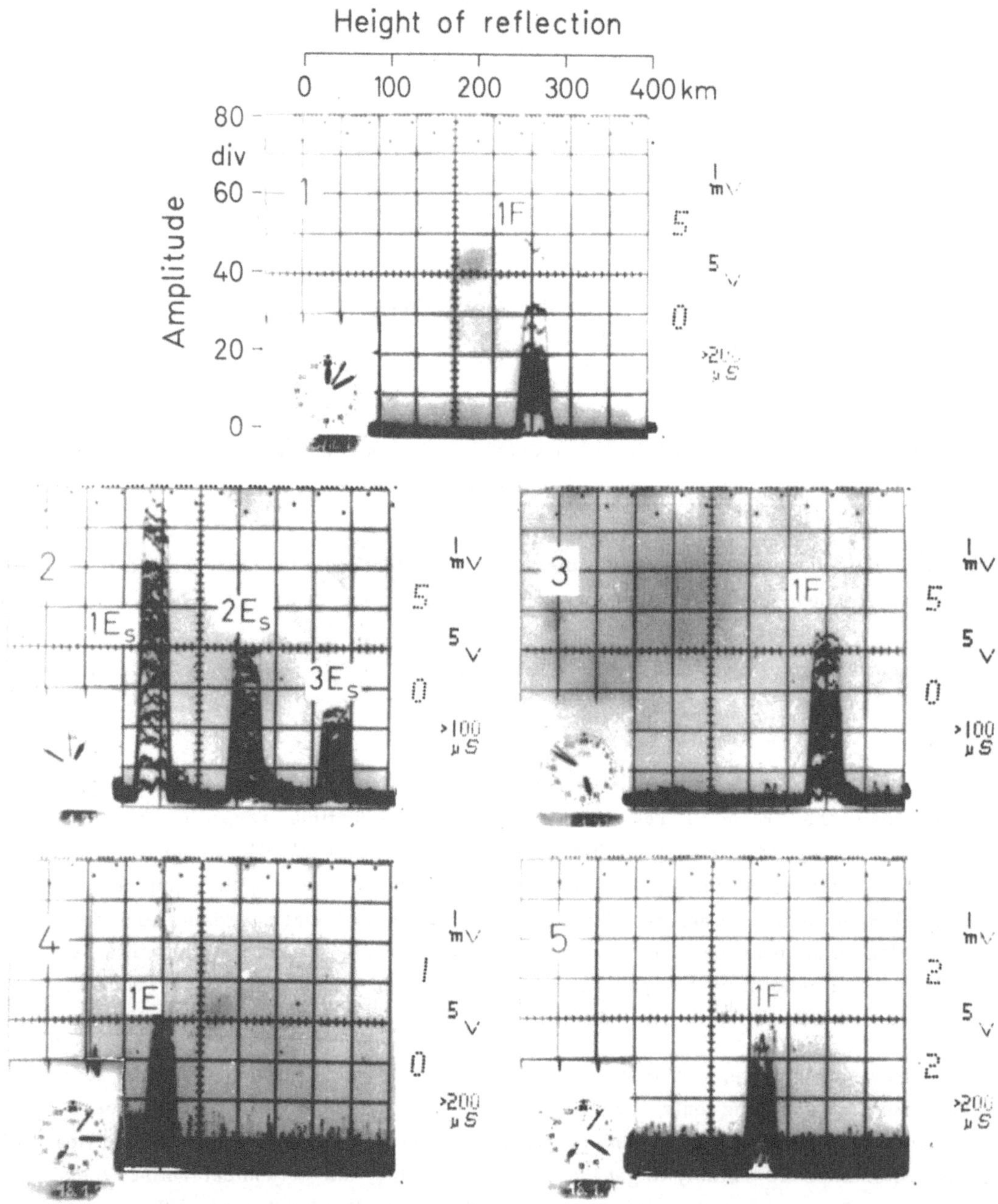

Fig. 3 : Some examples of pictures for demonstration of the photographic pulse-superimposing method showing different types of reflection from the ionosphere at various times and two frequencies. The time indicated by the clock is always given as UT.
3.1 A typical picture showing an 1 F echo obtained in the night of 15 to 16 January 1972. Latitude: 46.5°N, longitude: 16°W.

The period in which the picture was taken was from 00 : 10 to 00 : 20 UT. Some single pulses can be recognized above and below the mean pulse which has an average amplitude of 14 scale divisions; the scale is linear. The height is indicated at the top of the picture by a schedule of dots: The fourth line from above indicates the heights 0, 100, 200, 300, 400 km, the third line 50, 150, 250, 350 km, the second line 25, 75, 125 km, and so on, the first line 5, 10, 15, 20, 25, 30 km, and so on.
The code on the right means: Sensitivity of channel 1 (echoes) of the oscilloscope 1mV/div, attenuator 50 dB, vertical deflection factor of channel 2 (height marks) 5V/div, frequency used 1.73 MHz (0; 2 means 5 MHz); time-base sweep range > 200 μs/div. All these data are recorded automatically on each picture.
3.2 Multiple reflections from an Es layer (1Es, 2Es, 3Es) recorded just after sunset in the night from 31 December 1971 to 1 January 1972, 01 : 50 to 02 : 00 UT. Latitude: 10°N, longitude: 83°W. Pictures of this type were used to check the accuracy of the height measurements (see the included table), and also to determine the calibration constant. Moreover, the reflection coefficient of the Es layer can be estimated.

Echo	Virtual height of reflection (km)	maximum Amplitude (mm on the projected picture)	median Amplitude (mm on the projected picture)
1 Es	92	180	135
2 Es	184	100	60
3 Es	276	60	35

3.3 Reflection from the F layer (1 F) just before sunrise in the night from 31 December 1971 to 1 January 1972, 10 : 50 to 11 : 00 UT. Latitude: 10°N, longitude: 83°W (compare picture 3.2).
3.4 Reflection from the E layer (1 E) in the daytime; latitude: 51°N, longitude: 4°W.
3.5 Reflection from the F layer (1 F) in the daytime; f = 5.0 MHz.

The median values of the amplitudes can be found in different ways depending on the feature of the average pulse. It must be emphasized here that using a linear instead of a logarithmic scale is an important point; thus, the highest possible accuracy can be obtained. It should be mentioned that, if there is interference, recording of an "average pulse" instead of single pulses each on one frame enables to determine the median value of amplitude, and particularly the height of reflection, more precisely. This point was important because, due to the particular position of the derricks, during one passage, the transmitted pulse on the A-scope of the receiver was considerably out of shape thus disturbing very weak echos. By recording, however, of the complete diurnal variation of absorption, only a limited inaccuracy had to be taken into account, mainly at noon, and data L $(\cos\chi = 0.2)$ could be determined without severe difficulties.

4. Behaviour of ionospheric absorption on routes between 53°N, 10°E and 10°N, 83°W in winter 1971/1972

The behaviour of ionospheric absorption can be described most reliably and completely by characteristic parameters obtained from the diurnal variation. Thus, "the type of the day" can be recognized which varies considerably in winter giving much more information than single values as, for instance, noon values.

It is well-known [SCHWENTEK, 1958] that the diurnal variation of absorption L can be expressed by

$$L(f, t) = L_o \cos^n \chi (t + \tau), \qquad (2)$$

where f frequency, t time, L_o subsolar absorption, χ solar zenith angle, τ delay time. Measuring L(f, t) the values L_o, n, τ and L $(\cos\chi = \text{const})$ can be determined as the characteristic parameters, and used for comparison with the corresponding data obtained at other stations.

Fig. 4: Absorption L(1.73 MHz ; cos χ = 0.2) as a function of geographic latitude.

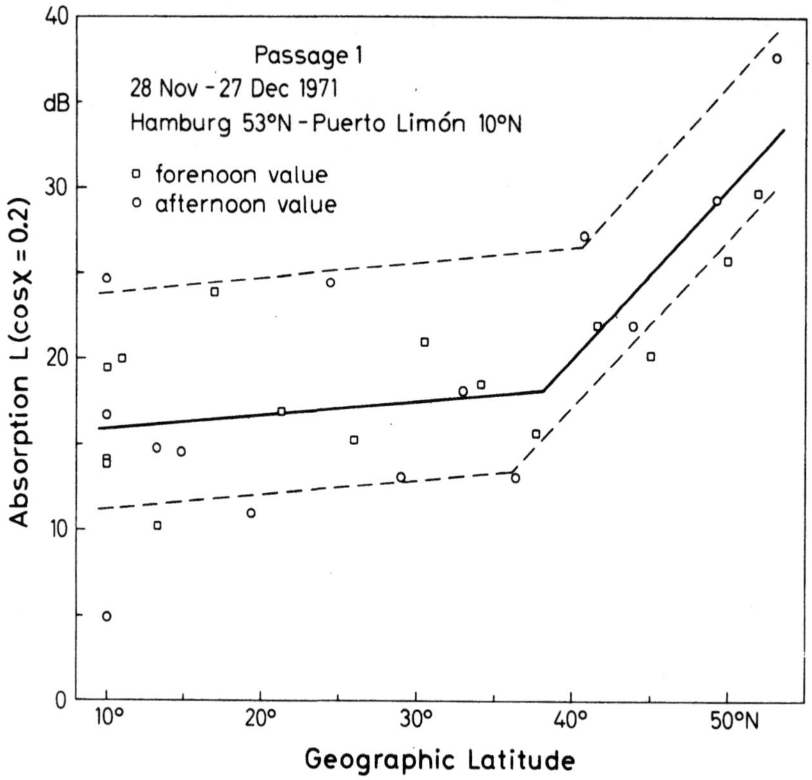

Fig. 4a

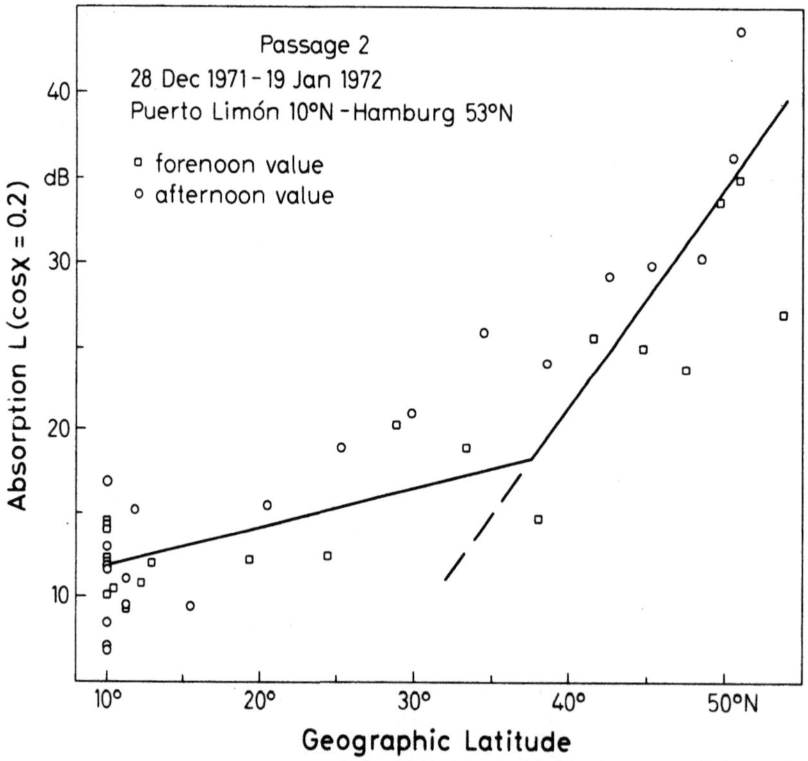

Fig. 4b

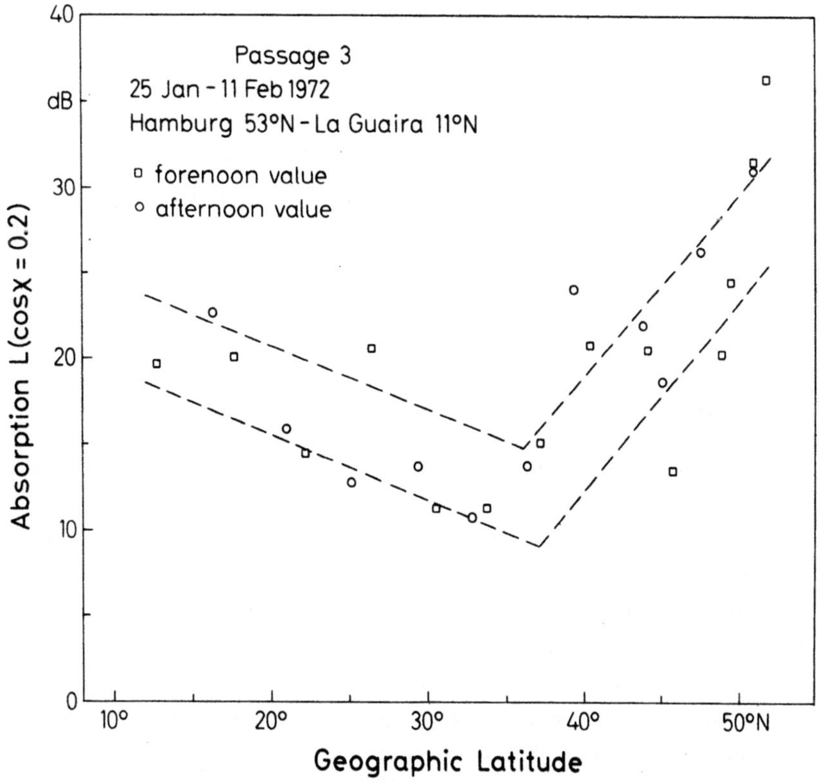
Passage 3
25 Jan - 11 Feb 1972
Hamburg 53°N - La Guaira 11°N
forenoon value
afternoon value
Absorption L(cosX = 0.2)
dB
40
30
20
10
0
10°
20°
30°
40°
50°N
Geographic Latitude

Fig. 4 c

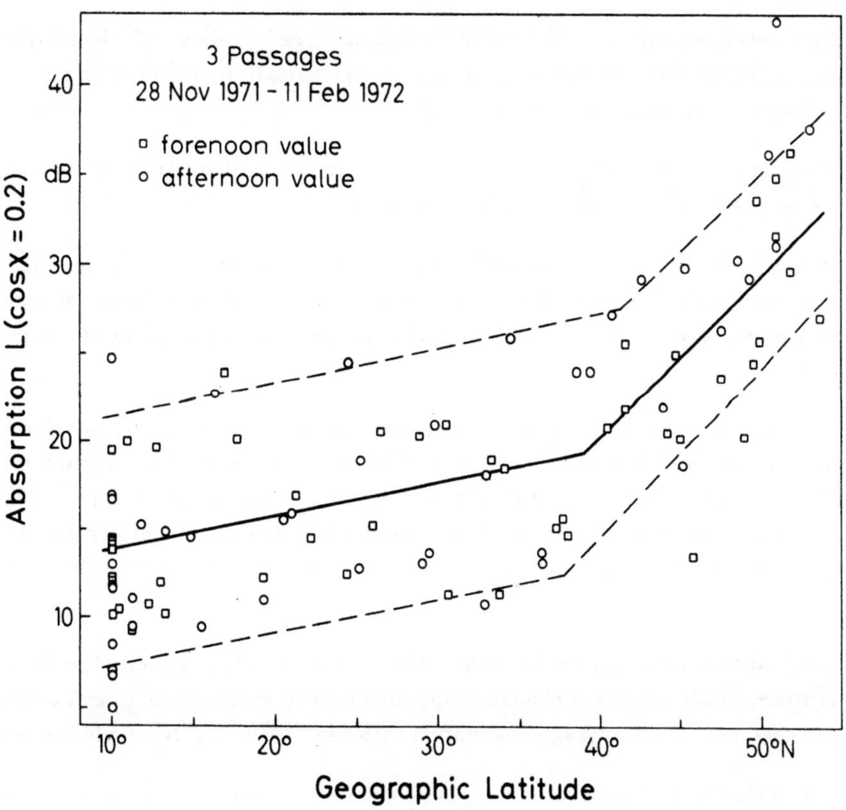
3 Passages
28 Nov 1971 - 11 Feb 1972
forenoon value
afternoon value
Absorption L(cosX = 0.2)
dB
40
30
20
10
10°
20°
30°
40°
50°N
Geographic Latitude

Fig. 4 d

A measurement on a moving ship might involve a small inaccuracy because of the variation of the position in the course of a day, in our case about 350 km in 11 hours. This point should be kept in mind considering the data.

While, during the night, it was sufficient to take one picture every ten minutes, averaging 60 pulses, in the daytime an average of 60 pulses was recorded within 2 minutes in order to comprehend also rapid changes in absorption. Moreover, in the daytime, we used always 3 steps of the attenuator differing by 10 dB (for instance 10, 20, 30 dB) to enable a selection of the optimal picture, that is, with an average amplitude in the center or upper half of the frame; in most of the cases even two pictures could be used. Thus, a rather large number of data was available for plotting of the diurnal variation of absorption $L = L(f, t)$. From these plots data $\cos\chi = 0.2$ were determined. These plots present a survey on the character of the day giving hints concerning the occurrence of sudden ionospheric disturbances, the behaviour of absorption around sunrise and sunset in the D region, and irregular periods in the diurnal variation. Moreover, $h'(f, t)$ was plotted for critical examination of the absorption values.

The characteristic parameters are used, at first, to describe the behaviour of absorption during the different passages. In Fig. 4 absorption at a constant solar zenith angle $\chi = 78.5^\circ$ ($\cos\chi = 0.2$) is plotted as a function of geographic latitude for each of the three passages. The forenoon and afternoon values were plotted separately because very often there was a considerable asymmetry in the diurnal variation. From Fig. 4 we may conclude that the behaviour of absorption at each passage is different although it shows a common trend. At the first passage a zone of transit can be recognized as the range between 35° and 41°N for, approaching from the North, at 35°N a change in the trend cannot yet be detected, and, on the other side, approaching from the South, a change in the trend does not occur earlier than at 41°N. This becomes evident if the points either below or above about 39°N are masked.

During the second passage the transit can be defined rather clearly; the change in the general trend occurs at about 38°N. The scatter of the data is smaller than during the first passage.

At the third passage again a well defined transit occurred at about 37°N although the behaviour at lower latitudes was different. Apparently, the day-to-day variation of absorption also at latitudes outside the zone of the winter anomaly can be rather high.

Taking the data of all passages together a description of a mean trend can be given showing roughly a transit zone between 34° and 44°N centered at 39°N.

It seems that a southern border could be fixed at 39° (at least, at longitudes between 20° and 35°W) but that it is not very well defined; perhaps it is more or less masked by local processes causing the large scatter in the absorption data. However, it is probably more fitting to speak of a transit zone than of a border.

A short note should be introduced here concerning the data and its interpretation. The scatter in the characteristic data is mainly due to variations in the ionosphere because the accuracy of the data is of the order of $\pm$ 2 dB. As a first step, all data obtained from the measurements are presented in the figures in order to show the real results independent from any theory. The curves drawn should be considered as a first approach of interpretation indicating trends only. It is clear that also other views are possible (see conclusions).

Another typical parameter, subsolar absorption $L(\cos\chi = 1)$, is plotted in Fig. 5 as a function of geographic latitude. It also shows a change in the mean trend at about 39°N and a transit zone from 35°N to 41°N. The constancy of the average absorption between 10° to 39°N should be noted.

Finally, the behaviour of the exponent n was studied; as expected n is a function of geographic latitude (not shown); and, in a similar way, also of magnetic dip angle (Fig. 6). Despite the large scatter due to ionospheric variations of those points which were obtained from the measurement on board the ship, a trend can be recognized which points to the southern border of the winter anomaly (dip 57°N).

In practice, a presentation of absorption vs. geographical coordinates (see Fig. 4d) would be more convenient, for instance , if maps should be used on board a ship. On the other hand, absorption can be plotted as a function of geomagnetic coordinates, for instance, in Fig. 7 absorption L (1.73 MHz; cos χ = 0.2) obtained from all three passages is plotted vs. magnetic dip angle. The lines show approximately the trend of the median values which are not included. It is true that a steady curve would fit the data somewhat better. However, thinking in trends, also in this case a zone of transit can be defined being centered at a magnetic dip angle of about 57° which corresponds to a geographic latitude of about 38°N.

With regard to these results it should be pointed out that absorption must be a real function of geographic latitude due to the formation of the D and E layer; but it is additionally a function of geomagnetic dip angle since the magnetic field acts on the produced ions and electrons if they are moved, for instance, in a circulation system. But from our measurements the degree of dependence on geographic latitude and geomagnetic dip angle cannot be separated.

In general, the results obtained from the measurements made at 5.06 MHz support those made at 1.73 MHz. In the daytime, the pulses of 1.73 MHz were reflected at the E layer, the pulses of 5.06 MHz at the F layer. Therefore, the influence of the particular conditions below the E-layer maximum, and at the level of reflection in the E layer, on the amount of the winter anomaly might be important only for 1.73 MHz, but not for 5.06 MHz, as far as an influence on the process of reflection is concerned.

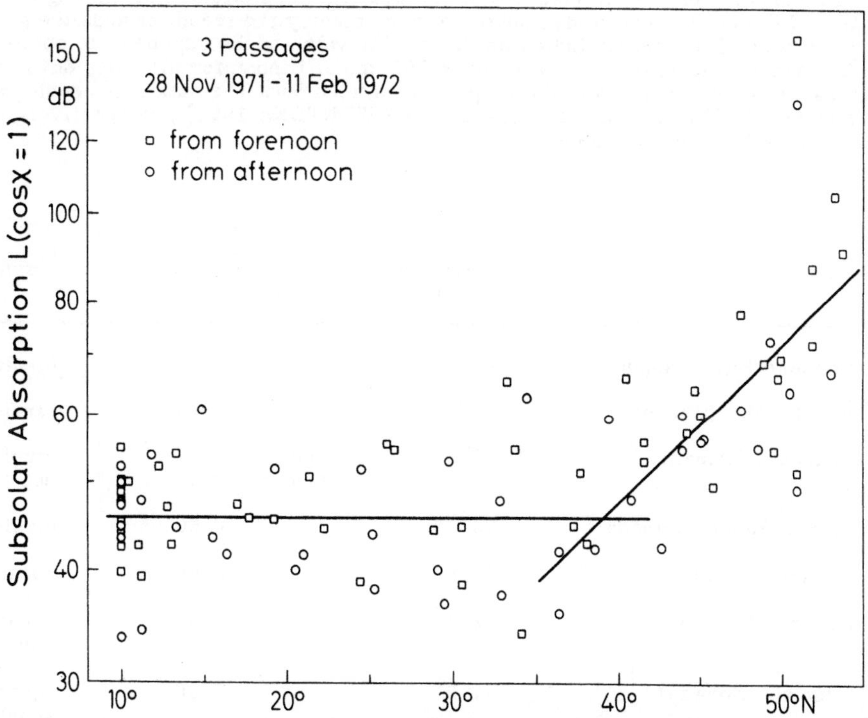

Fig. 5: Subsolar absorption L (1.73 MHz; cos χ = 1) as a function of geographic latitude obtained from measurement of the diurnal variation of absorption during the three passages of the ship.

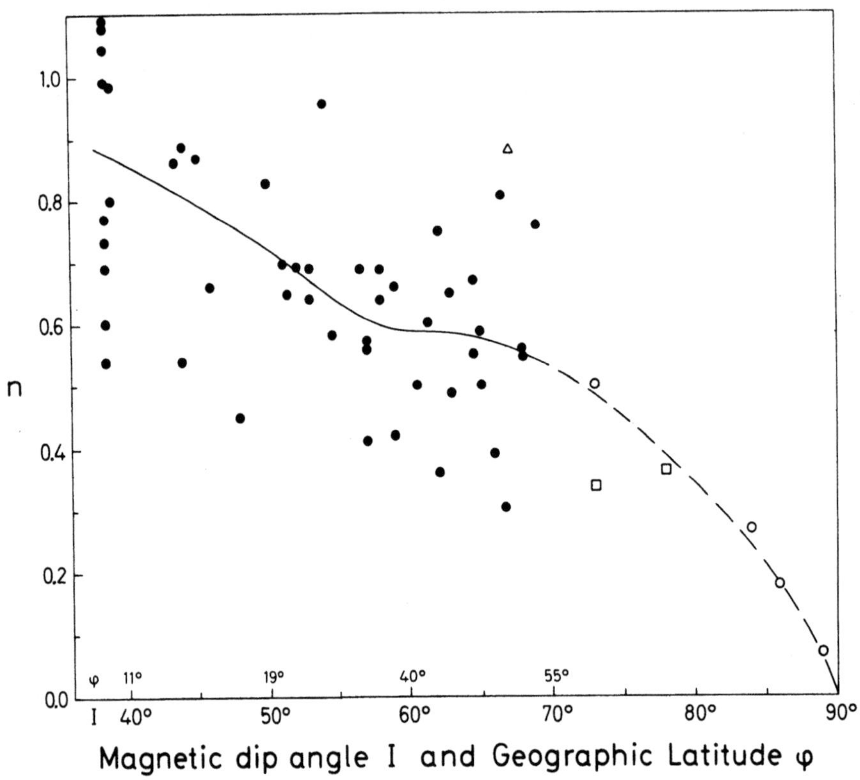

Fig. 6 : Exponent n (mean of forenoon and afternoon value) as a function of the geomagnetic dip angle (f = 1.7 MHz); the curve fits the overlapping median values. Some mean values of n obtained at various locations situated at dip angles greater than 70° are included to demonstrate more clearly the trend towards the geomagnetic pole. The triangle indicates the median value of the exponent n obtained from A3 measurements on 2.61 MHz over 300 km at Lindau for all those days on which measurements on board of the ship were carried out. The data represented as open circles [DAVIES, 1965] or squares [SCHWENTEK, 1968], respectively, are listed in the following table.

Location	geomagnetic dip angle		exponent n	symbol
Resolute Bay (Canada)	89°		0.07	circle
Baker Lake (Canada)	86°		0.18	circle
Churchill (Canada)	83°	Equinox	0.27	circle
		Winter	0.35	not marked
Prince Rupert (Canada)	73°		0.50	circle
Kjeller (Norway)	72.3°		0.60	not marked
Kjeller (Norway)	72.3°		0.34	square (from seasonal variation)
Tromsö (Norway)	77.9°		0.36	square (from seasonal variation)

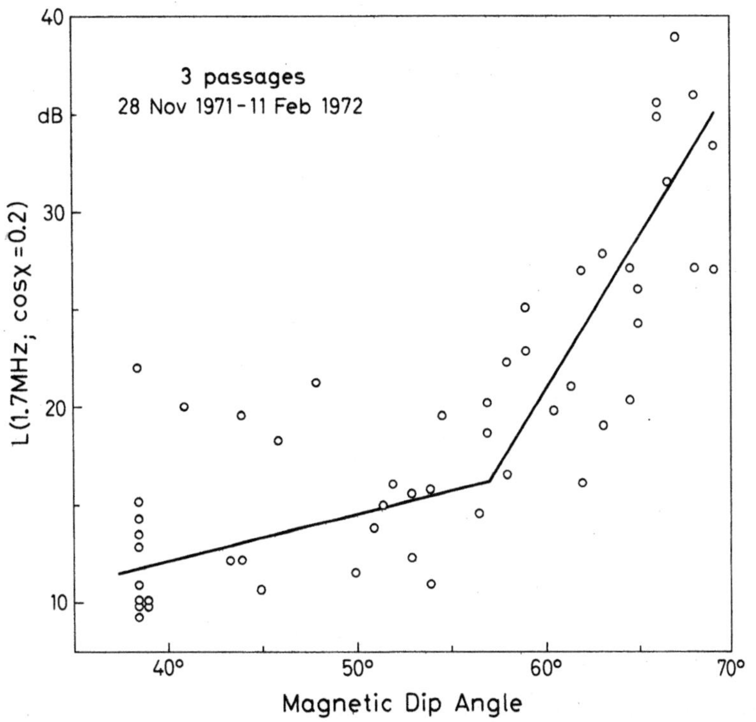

Fig. 7: Absorption L (1.7 MHz; cos χ = 0.2) as a function of the geomagnetic dip angle; the dip angle 57° corresponds to a geographic latitude of 38°N (see the route of the ship, Fig. 2).

Fig. 8 shows subsolar absorption L (5.06 MHz; cos χ = 1), as a function of geographic latitude; comparing it with Fig. 5 nearly the same feature reveals. From 10°N to 39°N absorption is approximately constant, from 39°N to 54°N, however, it increases considerably. A southern border of the winter anomaly zone can be presumed to be at about 39°N. A similar result revealed for absorption at a constant solar zenith angle χ = 78.5°, L (5.06 MHz; cos χ = 0.2), Fig. 9; trends and scatter of the points behave as for L (1.7 MHz; cos χ = 0.2).

Less clear, however, is the behaviour of exponent n in L (5 MHz) = $L_0 \cos^n \chi$. The scatter of the values is rather large (n = 0.55 to 1.35) and irregular. Roughly, the median values determined for overlapping ranges of geographic latitude are constant, n being approximately 0.85 ± 0.1 in the range from 10°N to 55°N; perhaps a weak minimum exists around 34°N, but this may be accidental (no Figure).

4.

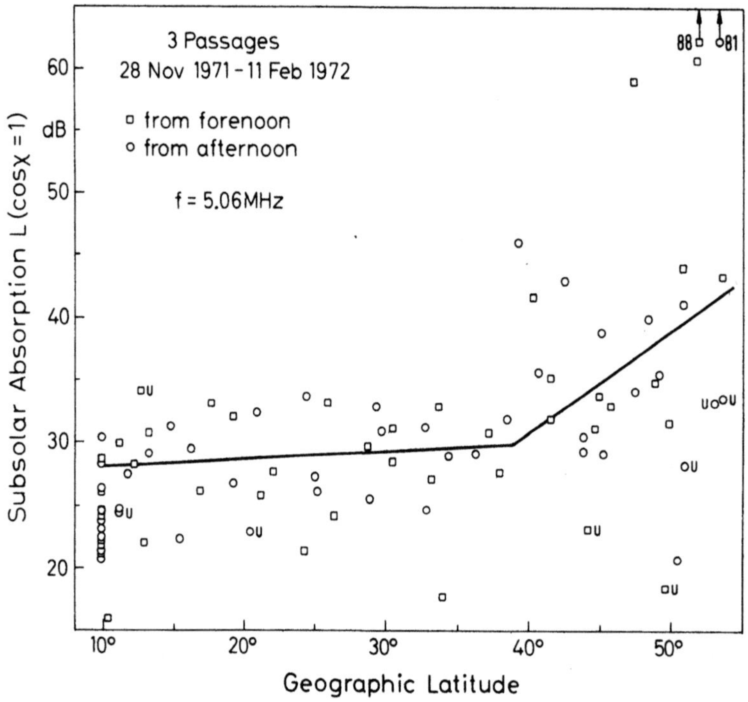

Fig. 8: Subsolar absorption L (5.06 MHz; cos χ = 1) as a function of geographic latitude. The line from 10° to 39°N approximately fits the median values, the line from 39° to 54°N is drawn rather tentatively to indicate the trend. U unreliable value.

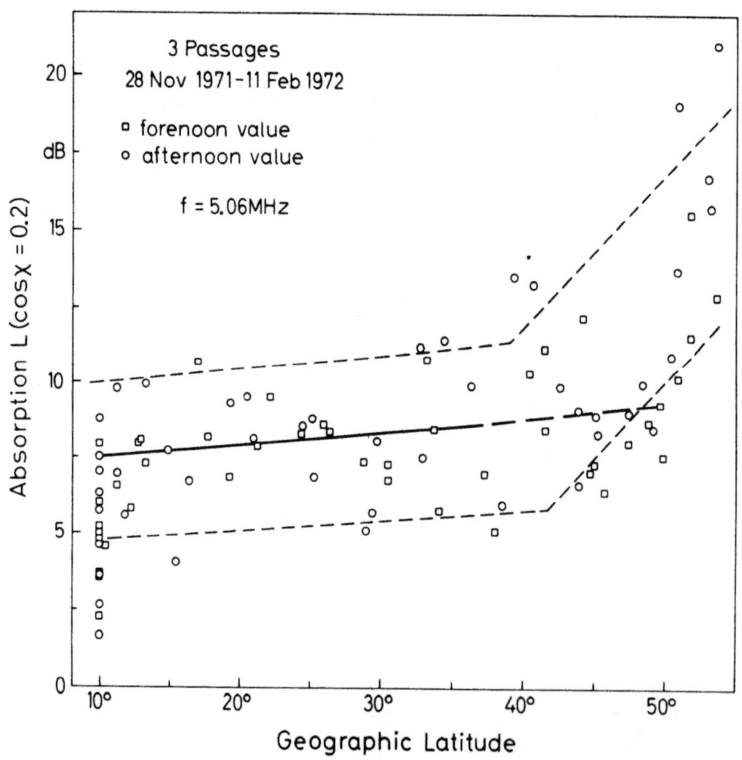

Fig. 9: Absorption L (5.06 MHz; cos χ = 0.2) as a function of geographic latitude. The lines are drawn tentatively to indicate the trends.

5. Comparison of data obtained on board the ship and, at the same time, at fixed stations

During the entire period of the expedition the circuit Norddeich-Lindau (A3; 296 km; 2.61 MHz; 52.63°N) was used as a reference. This circuit is in a geographic position where a strong winter anomaly exists. On board, we received daily via Norddeich-Radio the values of L (cos χ = 0.2) obtained from the circuit Norddeich-Lindau.

The day-to-day variation of L (cos χ = 0.2) determined at Lindau and on board the ship is plotted in Fig. 10 separated for forenoon and afternoon values. At passage A the correlation between the absorption values can be pursued until December 7 (position of the ship 40.7° N). The next day, December 8, the absorption measured on board the ship was approximately the same as on December 7; the circuit Norddeich-Lindau, however, showed a day of excessive absorption. In this case, we may conclude that between latitudes of 40.7° to 37.7°N at least the correlation vanished away, and also the strong winter anomaly. We cannot conclude with certainty that the disappearance of the correlation indicates that we had reached already the border of the winter anomaly zone. For it is known

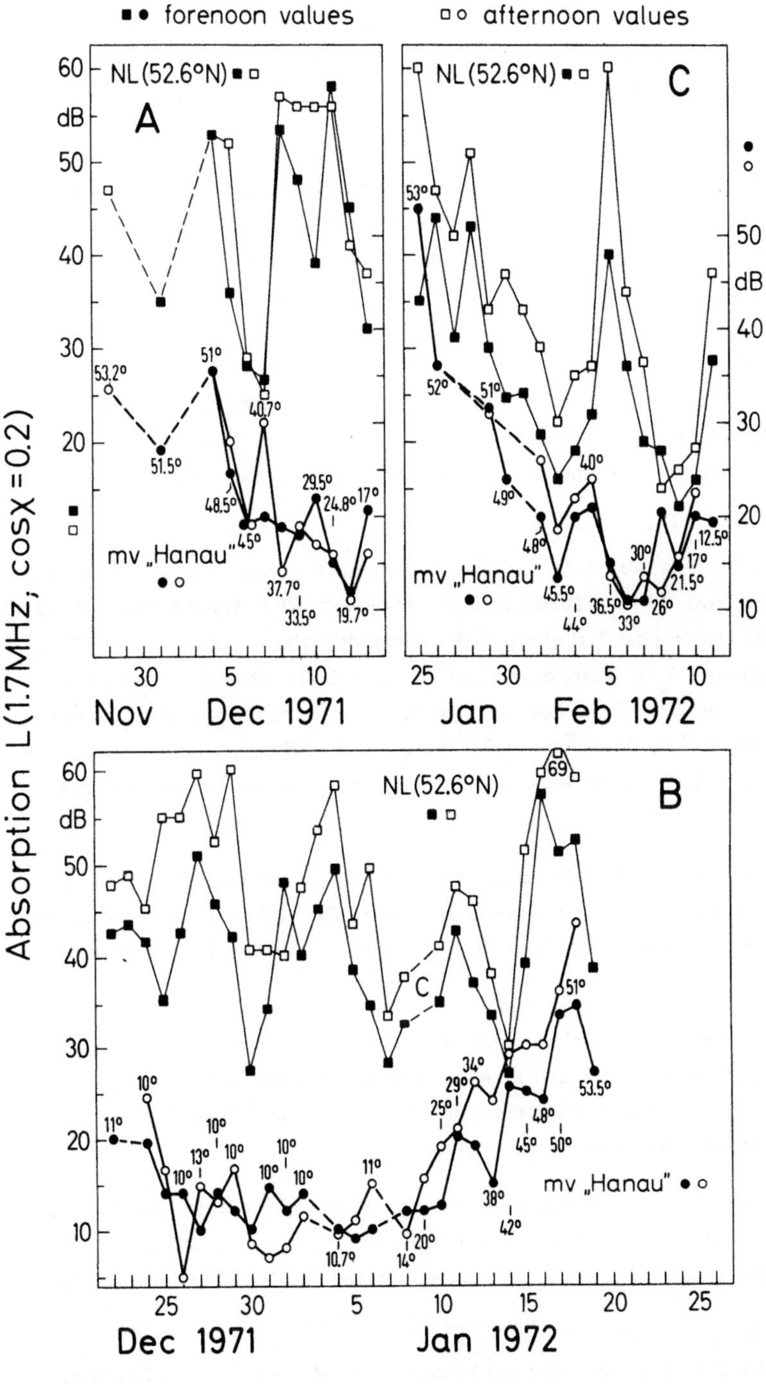

Fig. 10: Absorption L (1.73 MHz; cos χ = 0.2) at a constant solar zenith angle χ = 78.5° determined from plots of the diurnal variation of absorption (first passage), and from plots of log L vs. -log cos χ (second and third passage), respectively. - For the first passage absorption L (1.73 MHz; cos χ = 0.2) obtained from plots of log L vs. -log cos χ is not presented here because it behaves rather similar as the data shown in the figure, and because in the period from 28 November to 4 December inclusive during very short periods of the diurnal variation only the absorption could be measured, either in the morning or afternoon during a few hours. - The latitude values in the figure indicate the ship's noon position at each day.

the amount of the winter anomaly to decrease with latitude [DIEMINGER et al., 1966 ; SCHWENTEK, 1968; WAKAI et al., 1970] . And although the absorption at 37° N was rather low at the time of measurement other days might exist on which the winter anomaly is present to some extent (see Fig. 1). During passage C the correlation holds until February 4. On February 5, however, the absorption measured on board the ship decreased by about 5 dB, whereas the absorption at the circuit Norddeich-Lindau increased excessively by about 20 dB. Again, between 40° and 36.5° N, the correlation disappeared. At February 8 and 9, the absorption at 52.3° N was as low as at 21° N; on these two days the rare case happened that no winter anomaly occurred at Lindau (see Fig. 1).

Finally, passage B shows that at about 10° N absorption remained approximately constant while at 52.6° N large variations occurred, the afternoon values generally being greater than the forenoon values. In this case, the start of a correlation cannot be stated clearly.

The general problem arises here, how the correlation between the occurrence of excessive absorption at different stations might vary with distance and time between the stations, where, probably, the variation with distance is different dependent on whether it is a variation in latitude or longitude. Moreover, the correlation might also be different from period to period. This problem can be studied most effectively with a rather dense network of stations, where the same operating frequencies, standardized equipment, and data processing are used. A first step in this way was done by the European Cooperation in winter 1971/72. Additionally to the circuit Norddeich-Lindau the circuit Balerma-Aranjuez was operated at Spain (f = 2.83 MHz ; method A3 ; point of reflection at 38.4° N) by our institute ; moreover, absorption measurements were made simultaneously at Lancaster (method A1 ; f = 2.14 MHz). Thus, a first comparison was possible of the behaviour of absorption at those stations with regard to the absorption measured on board the ship.

The values $L(\cos \chi = 0.2)$ obtained from the circuits Norddeich-Lindau (NL) and Aranjuez-Balerma (AB), on board the ship, and noon values from A1-measurements at Lancaster (L), show a similar trend until December 6 inclusive (Fig. 11). On December 7 there is a discrepancy, since absorption at NL still decreased considerably. From December 7 to 8 absorption at NL and L increased excessively, and also at AB the increase was remarkable ; the ship, however, seems just to have moved out of the zone where the anomaly occurred(from 40.75° N, 22.35° W to 37° N, 30.5° W). For, at 40° N the station nearest to the ship was the circuit AB, and it should be expected the data of these two stations to correlate best.

Later on, NL, L, and AB show a rather high level of absorption, while at the ship absorption is rather low. Whether there is a real correlation between AB and the ship from December 8 to 12 is an unsolved question; we suppose that AB correlates with NL (see Fig. 1), and that the similar trend of absorption at the ship might be accidental.

During the second passage a different behaviour of absorption was found at the four stations. NL and L showed a maximum of absorption around January 11, and a very large one around January 17. The first period shows a similar form also at AB ; the second period, however, does not show the large increase of absorption as observed at the higher latitudes.

The interpretation of absorption measured during the second passage on board the ship is difficult. Our version is that the values obtained at AB and on the ship correlate from January 12 or 13 (when the ship reached 38° N) to 16, and that from January 17 to 19 the values obtained on the ship followed the behaviour of absorption at NL and L.

A general conclusion from these results is that probably the behaviour of absorption as a function of geographic position is due to very complex processes occurring at the heights between 85 and 110 km, processes which perhaps are associated to variations in the large-scale wind pattern.

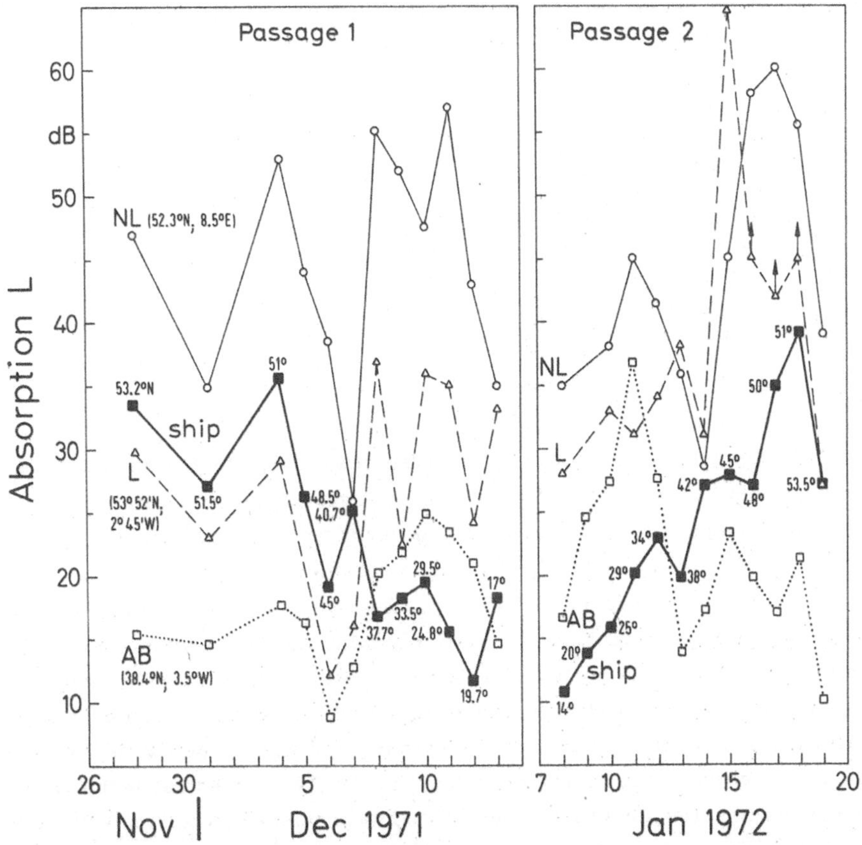

Fig. 11: Absorption L measured simultaneously at three fixed stations and on board mv "Hanau". NL circuit Norddeich-Lindau (2.61 MHz; 296 km; A3; equivalent frequency 1.5 MHz); ship (A1; 1.73 MHz); L Lancaster (England; A1; 2.14 MHz); AB circuit Aranjuez-Balerma (Spain; 2.83 MHz; 374 km; A3; equivalent frequency 1.34 MHz).

6. A note from the view of theory

The first interest of our study was to pursue the real variations of absorption from day to day; another view is to consider the average behaviour. Both are related as weather and climate. The average behaviour was studied as follows. Noon absorption derived from the diurnal variation (regression curves) was plotted vs. solar zenith angle χ for all days of the three passages. Thus, at first, the days showing winter anomaly could be recognized clearly. Drawing the regression line excluding those days the mean absorption value was determined for $\chi = 40^\circ$, as an example. This value, L (1.7 MHz; $\chi = 40^\circ$) was found to be 37.5 dB. It can be derived also from theory. MAEDA [1971] has compiled electron-density profiles N(h) measured by rockets for winter. An average exponential profile was derived from these single profiles for heights h from 70 to 100 km, and for $\chi = 40^\circ$, described by

$$N = N_o e^{ah} ,$$

where $N_o = 6.28 \times 10^{-7}$ el/cm^3, $a = 0.253$ km^{-1}. The collision-number profile [ELLING, 1968] was assumed to be

$$\nu = \nu_o e^{-bh} ,$$

where $\nu_o = 7.8 \times 10^{11}$ sec^{-1}, $b = 0.161$ km^{-1}.

Using these profiles, the complex refraction index for 1.7 MHz was computed as a function of height (Appleton-Hartree formula), and absorption derived from the phase integral

$$L = -k\,\mathrm{Im} \int_0^{z_o} n(z)\,dz \ ,$$

where k = constant.

At the complex height $z = z_o$, n is zero. Integration yielded L = 35.6 dB. Applying the full-wave theory [see GRUSCHWITZ, 1972] it was found L = 36 dB.

7. Discussion

One objective of our studies of which a part is presented in this paper is to obtain a global survey on the behaviour of absorption as a function of geographic and geomagnetic coordinates, and to use simultaneously absorption as a diagnostic indicator of the processes occurring in the D and E region. Absorption, however, is a complicated function of a number of parameters. The production of the D and E layer where absorption takes place depends on the radiation of various parts of the solar spectrum and its different variations in time, and the conditions of the neutral atmosphere, circulations included. While the neutral atmosphere can be studied as a function of geographic latitude and longitude, height above ground, solar zenith angle, diurnal, seasonal and solar cycle variations, the study of the ionized D and E region within the neutral atmosphere requests to take into account additionally the Earth's magnetic field. Experimental implications are that in the course of the day, at the frequencies below the critical frequency of the E layer, the height of reflection of the echos can vary considerably, particularly in winter at middle latitudes. This is most pronounced around noon, varying from day to day. Therefore, for comparison, noon values will not be as suitable as values for constant solar zenith angle when obtained from the diurnal variation of absorption, and separated for forenoon and afternoon. Recording the diurnal variation also solar flare effects can be easily detected and those periods excluded. At higher latitudes particle precipitation can produce additional absorption. Besides, the dependence of absorption L (cos χ = const) on sunspot number is a function of season, at least, at middle latitudes [SCHWENTEK, 1971] .

Therefore, it should be emphasized, that all these points have to be considered if absorption is intended to be used as a diagnostic indicator of the behaviour of the D and E region on a global scale. Since absorption also depends on frequency, it is urgent to measure, if synoptic purposes are in mind, preferably at vertical incidence, and on standardized frequencies, as, for instance, 2.2 MHz, and if possible also 1.7 MHz ; or even more. - The asymmetry in the diurnal variation of absorption which is a common feature [SCHWENTEK, 1968] is very much more pronounced at lower latitudes. At Lindau, the time of maximum of absorption is normally delayed by 0 to 30 minutes against the time of minimum of solar zenith angle, but at lower latitudes, the delay increased to about 120 to 180 minutes. A similar asymmetry in electron-density profiles (on which absorption depends) about noon of up to 6° in solar zenith angle, the morning electron densities being lower than those at the corresponding time in the afternoon, was reported by COYNE and BELROSE [1972] who made differential-absorption partial-reflection experiments at 2.66 and 6.275 MHz at Ottawa (45° N; 76° W). This asymmetry in electron-density profiles was also found by GRUSCHWITZ [1972] from MF multifrequency absorption measurements carried out at Tsumeb (19° S).

Finally, we should like to point out that the term "winter anomaly" is not restricted to ionospheric absorption only. This term refers evidently to every anomaly within the seasonal variation of a parameter. Indeed, at first, the term was used concerning the absorption of shortwaves in the ionosphere. The fact is that, at medium latitudes, absorption at a constant solar zenith angle, L (cos χ = const), shows considerable enhancements in winter, and also very large fluctuations. But other parameters of the atmosphere show a very similar behaviour, for instance, critical frequency of the E layer measured at cos χ = const. Moreover, also the stratospheric warmings must be considered on the background of a regular variation of stratospheric temperature with the sun's zenith angle at noon in the course of the year, and can be called winter anomaly in stratospheric temperature [LABITZKE and SCHWENTEK, 1968] .

The term "winter anomaly" seems to be applicable to the anomalous behaviour in winter of the stratosphere, the mesosphere, and the ionosphere, thus expressing a general deviation from a dependence of essential parameters from the solar zenith angle. However, considering the problem of seasonal anomalies not only the dependence on the Sun's zenith angle is important but also the dependence on geographic and geomagnetic latitude of the parameter. As an example, J.W. WRIGHT [1963] showed that the seasonal anomaly in F-region maximum electron density, and still clearer in electron content below the peak of the F2 layer, in fact, is not a winter but a summer anomaly.

Here also a result presented by WAKAI et al. [1970] should be reminded showing that the southern border of the winter-anomaly zone in Japan must exist between latitudes 37.5° N and 35.7° N. They mention that winter anomaly is sometimes observed at 35° N (Kokubunji), and occasionally even at 31° N (Uchinoura). Probably, also at longitudes of about 140° E there is a boundary at about 37° N; but winter anomaly conditions may sometimes extend more to the South, down to 31° N, perhaps systematically as a function of solar activity, or sporadicly due to variations in the wind system in the lower thermosphere.

It is remarkable that the latitude range 30 - 40° N seems to be a transit zone not only for the seasonal behaviour in ionospheric absorption but also for stratospheric and mesospheric temperature (that is, southward, there are no more stratospheric warmings but simultaneous small coolings; LABITZKE, 1972), and also for radiances emitted from the stratosphere in the center of the 15-μm CO_2 band as measured from the Nimbus 3 satellite [FRITZ and SOULES, 1972] . Thus, it appears that in the zone 30 to 40° N the change occurs from the lower to the median latitude region, not only in the troposphere but also in stratosphere, mesosphere, and ionosphere.

8. Conclusions

1. It was verified that the method of integrating a certain number of pulses by superimposing them on one picture (24 mm x 24 mm) is very useful to obtain quickly median values of absorption, particularly, on expeditions as, for instance, on board a ship. This method allows also to measure the height of reflection, and to take into consideration the shape of the pulses. The data processing can be done rather simple; it is possible to plot manually the diurnal variation of absorption with a delay of one day only. Also the night-time values can thus be determined. In this way, reliability and constancy of the whole circuit, the antenna inclusive, can be checked day by day. This is important during a passage on board a ship to have a current control, and to avoid failures.

It was realized that the use of an electronic desk calculator with extended memory and a plotter made possible a very quick and simple data processing already during the passages. The data obtained from the frames, as time, attenuation, height of reflection, median values of the amplitudes of the pulses, and the position of the ship were put immediately into the calculator. The plotter drew the L vs UT curve and the plot log L vs - log cos χ , including regression lines, separated for forenoon and afternoon, and printed the data characterizing these lines, as exponent n, subsolar absorption L (cos χ = 1), and absorption at cos χ = 0.2 .

2. Operating the equipment on board a ship the main problem was to install the antenna. The type of the ship used was rather suitable concerning the dimensions; but there was some trouble due to the derricks. In a certain position the width of the received direct pulse from the transmitter indicated on the A-scope was

broadened up to h' = 100 km; this effect was probably due to damped oscillations emitted from the derricks acting as an "after-sounding system". Therefore, before starting the passage, the derricks should be put into the optimal position.

3. The behaviour of absorption was somewhat different at each passage. But considering all passages now available, including those of LAUTER and SCHÄNING [1970], not only a similar trend in absorption as a function of geographic latitude was found but also the striking result that something like a southern border of the winter anomaly could be determined to exist at about 37° N. On the other side, the continuous measurements made on the circuit Aranjuez-Balerma (Spain) at a reflection point of 38.4^{o} N showed a pronounced winter anomaly in absorption at 2.83 MHz, at least, during December and January. Therefore, for future work, a greater number of passages should be still undertaken, at different times in winter, and in different winters in the solar cycle, but always in cooperation with fixed, carefully placed stations on land or islands between 60° and 30° N. Thus, the variation in time of the "absorption pattern" within a larger region could be studied most effectively as a function of both geographic and geomagnetic coordinates. It should be noted that, geographically, the rocket launching site Wallops Island is located at approximately the same latitude as the circuit Aranjuez-Balerma in Spain. Therefore, it is probable that at Wallops Island as in Spain pronounced winter anomaly will occur in December and January only; longitude and dip, however, are rather different and might modify the behaviour to some extent.

II. An equipment for an A1 mobile station

In section I some essentials of the antenna and the equipment have already been outlined; in this part a more detailed description will be given.

Fig. 12 shows a block diagram of the equipment without power supplies. All parts are installed in various slide-in units which were mounted into 19"-double cabinets; the height of each cabinet is 1.4 m, the width 1.1 m, the depth 0.65 m (Fig. 13).

1. Pulse modulator

A hard-tube pulser was used as a pulse modulator. In contrast to the radar technique in this modulator a 1:1-pulse transformer is applied. Thus, the cathodes of the modulator and transmitter tubes could be connected to ground; moreover, the pulse-voltages supplied to the screengrids and anodes of the transmitter valves could easily be obtained from taps at the secondary of the pulse-transformer. -

In the modulator two switching tubes are used in parallel. The dc-energy for the modulator is accumulated in a 4.4 μF condenser at 12 kilovolts. Thus, the modulator can produce power pulses of 10 kilovolts and 15 A at the secondary of the pulse transformer.

The pulse length can be selected between 50 to 250 μs, the repetition frequency in the range from 10 pulses/s to 0.1 pulse/s. However, the greatest possible product of the pulse length times repetition frequency is restricted not to be greater than 10^{-3}.

The control grids of the switching tubes are biased at -500 volts and driven by a bootstrap driver with a pulse-amplitude of 500 volts. This driving gridpulse can be applied with a flat or with a continuously adjustable rising top, so that the declining top of the power pulse (as caused by the pulse transformer and the drop of the voltage of the 4.4 μF storage condenser during the time of pulsing) can be compensated. -

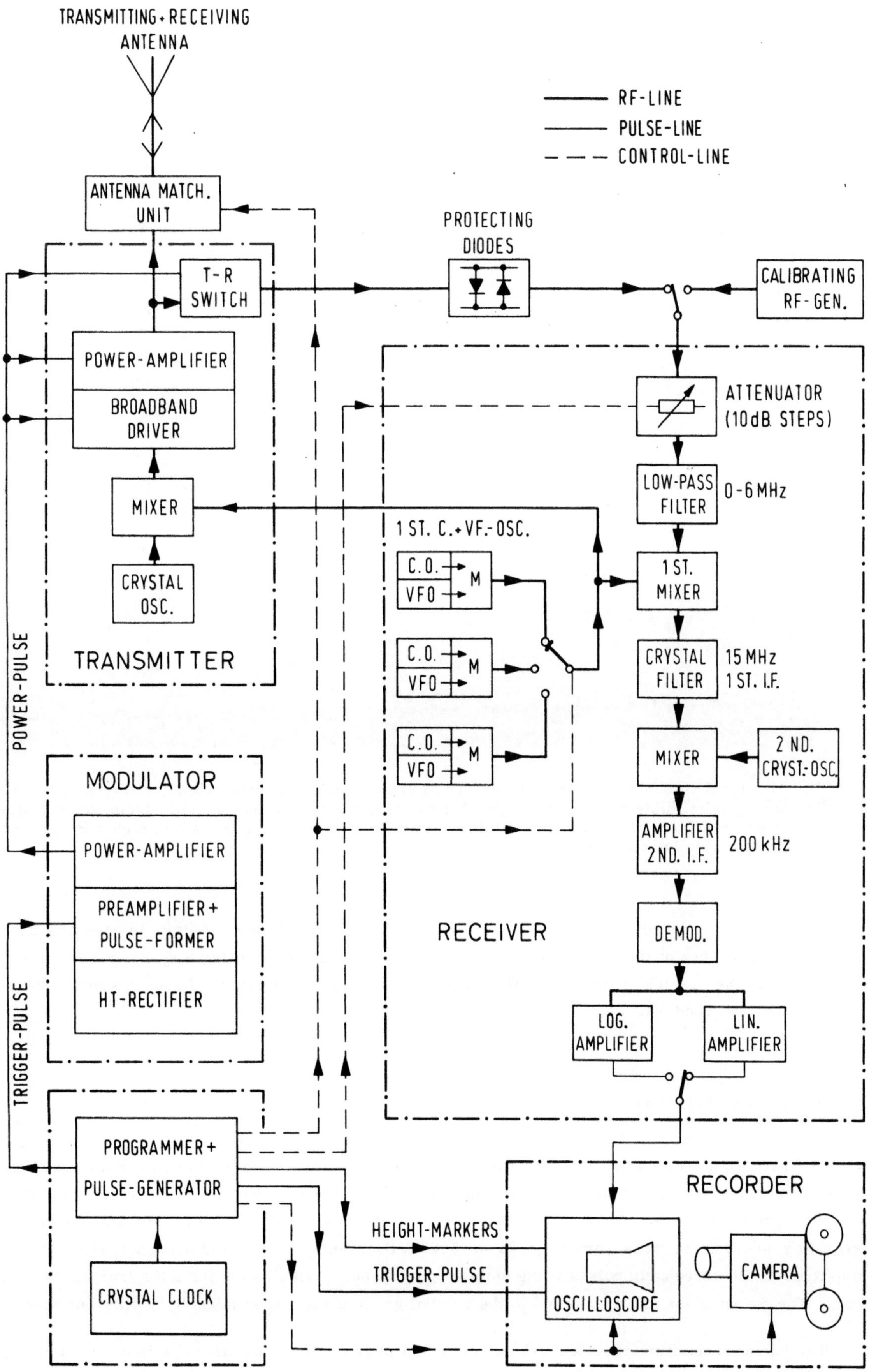

Fig. 12 : Block diagram of the A1 mobile equipment used on board mv "Hanau".

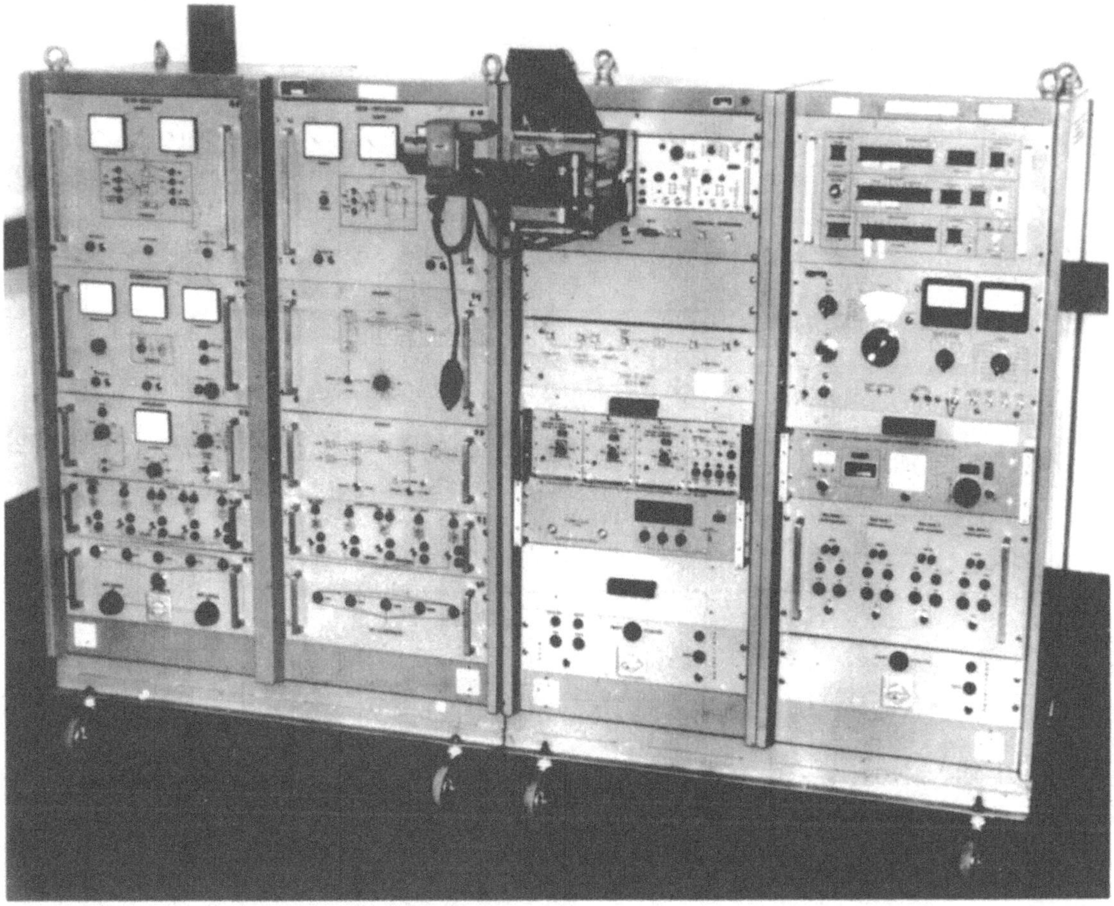

Fig. 13 : Total view of the entire A1 mobile equipment (two separate cabinets). From left to right: Cabinet 1, left: Pulse modulator; right: Pulse transmitter. Cabinet 2, left: from top to bottom: Recorder, receiver, three local oscillators, remotely controlled attenuator; right: Programmer, signal generator for calibration, crystal clock.

The storage condenser of 4.4 μF is charged by a one-way silicon-rectifier. In case of a flash over in the switching tubes of the modulator or the power tubes of the transmitter a spark-gap is immediately fired and thus the 4.4 μF-condensor will be discharged within some tenths of a μs; simultaneously, the rectifier was switched off.

2. Pulse transmitter

At first, the power stage of the transmitter consisted of two pulse tetrodes in parallel, and of a π-network in the anode circuit for matching the impedance of the tubes to the 50 ohms coaxial cable feeding the antenna. The condensers and the coil of the π-network were switched by the programmer according to the frequency used for the measurement (1.73 or 5.06 MHz). Both, anodes and screengrids, were pulsed, the amplitude of the screengrid pulses being stabilized by Zener diodes maintaining the transmitter output constant. The output of the driver was fed to the control grids of the power tubes by a wide-band transformer.

The pulse transmitter provided r-f pulses with a peak power of 60 kilowatts into 50 ohms in the frequency range from 1 to 6 MHz. Concerning pulse length and repetition frequency see section 1, pulse modulator.

In order to make the transmitter useful also for frequencies below 1 MHz a push-pull power stage must be used due to severe limitations which are caused by use of filter chokes in the output stage. Such a wide-band power stage covering the frequency range from 250 kHz to 6 MHz was constructed after the expedition by ship to replace the tuned power stage described above.

The driver was designed and built as a wide-band unit. It consisted of two transistorized stages (class A amplifiers) followed by a pentode and a cathode follower which fed the last stage over a wide-band transformer. Anodes and screengrids were pulsed, the amplitude of the screengrids pulse being stabilized by Zener diódes.

Finally, by mixing the frequency of the first oscillator of the receiver with the first intermediate frequency the transmitter frequency was generated; thus, an exact tuning of the receiver to the transmitter was warranted.

A simple antenna-system for transmission and reception of signals was used on board the ship by employing a T-R switch consisting of ht silicon diodes operated by the power pulse of the modulator. Thus, the receiver was disconnected from the antenna only as long as the transmitter was pulsed.

During any other time a suitable dc current flew through the diodes putting them into a low impedance state; then, reception of ionospheric echoes was possible.

3. Programmer and pulse generator

The programmer was designed as a unit providing for various programs. From 10 kHz-pulses obtained from a crystal clock all the required trigger pulses and control functions were derived in the programmer.

The trigger pulses for the modulator and the recording oscilloscope had a length of 10 µs; its repetition frequency could be set by thumb-wheel switches in the range from 10 pulses per s to 1 pulse per 15 s. Height calibration marks were provided for every 100, 50, 25 and 5 km; their width on the A scope was continuously adjustable for a good reading on the record (see Fig. 3).

The shutter of the recording camera and, independently, the film transport were also triggered by the programmer, or could be operated manually; range: 1 per minute to 1 per 15 minutes.

The programmer was designed for control of a multi-frequency A1-equipment operating at a maximum of ten frequencies. The number of frequencies to be used could be determined arbitrarily, for instance two, 1.73 MHz and 5.06 MHz. In this case the frequencies were switched on, one after the other, each for a preselected time interval, the duration of which could be set by a thumb wheel in the range from 1 to 15 minutes. Simultaneously, corresponding to the frequency just switched on, the most suitable of 10 possible antennas, and the necessary position at the attenuator (between 0 and 90 dB in steps of 10 dB) for obtaining echoes of suitable amplitudes could be selected.

The entire circuitry was constructed by use of integrated circuits. It was screened by installing it into a metal box, and filtering very thoroughly all lines for input and output. Thus, interference by the strong pulses generated in the pulse modulator and the transmitter could be avoided.

4. Receiver and recorder

The ionospheric signal coming from the T-R switch was fed to an attenuator via a set of protecting diodes. The attenuator was switched by the programmer in 3 steps of 10 dB; therefore, the programmer had to be adjusted manually from time to time for optimal recording in order to prevent saturation of the receiver.

The receiver was of the double superheterodyne type, and manufactured in solid state technique. The rf signal from the attenuator output was fed directly to the first mixer, a diode-ring mixer, via a low-pass filter, the output of the mixer into a crystal filter tuned to 15 MHz as the first intermediate frequency. The crystal filter especially developed for pulse transfer (a bandwidth of 10, 20, or 40 kHz could be selected) was followed by the second mixer generating a second i-f of 200 kHz, an i-f amplifier of several stages with a bandwidth of 60 kHz, a demodulator, and a linear and also a logarithmical amplifier which could be selected as desired.

The different stages of the receiver were built as separated units and connected with each other by 50-ohms cables. In order to obtain an excellent pulse response the receiver was equipped with filters corrected for transit time.

Finally, some specifications of the receiver: Cross modulation stability of the first mixer better than 80 dB; image frequency rejection better than 75 dB; linear range greater than 60 dB with a maximum input signal of 10 millivolts; sensitivity 1.85 μvolts at a signal to noise ratio of 6 dB, and 6.5 μvolts at 20 dB, respectively; overall gain 25 dB (attenuator at 0 dB). -

The output of the receiver was fed to an oscilloscope (Tektronix type 5030) with an A-scope type display and photographed with a 35 mm automatic camera (Robot II); for details see Fig. 3. Movement of the shutter and film transport were provided by the programmer.

5. Antenna

The frequencies used on board the ship were 1.73 and 5.06 MHz. Their ratio is approximately 1:3. The most suitable antenna would have been a $\lambda/2$-folded dipole for 1.73 MHz. Such a dipole was impossible to mount because the distance between the masts was too short. Only a dipole of 2 x 34 m = 68 m instead of 87 m could be installed (Fig. 14). Its geometrical height above the deck was 26 m, and the height above the sea level approximately 31 m. But these dimensions are not so important because of the influences of the superstructure, a lot of wire-ropes, and the derricks.

Since the dipole was electrically too short, there was a strong capacitive frequency dependent component in the impedance of the antenna. Because of the heavy-weather conditions which had to be expected during the winter passages of the Atlantic Ocean it was decided that no tuning circuitry at the antenna terminals at the top of the signal mast should be installed. Therefore, only a simple guanella-transformer, symmetrical to unsymmetrical, was used for matching the antenna to the 50 ohms unsymmetrical feeding cable. That transformer was encapsulated completely waterproof in a plastic case which was filled up with transparent, high insolating silicon-resin (Fig. 15). The compensation of the reactive component and the transformation of the active component of the antenna took place at the transmitter end of the cable by an appropriate length of the cable and by coils which were switched for the different frequencies by high-tension relais, controlled by the programmer. Thus, a load of exactly 50 ohms was connected to the output of the transmitter and at every time full suspection of all tuning components was possible.

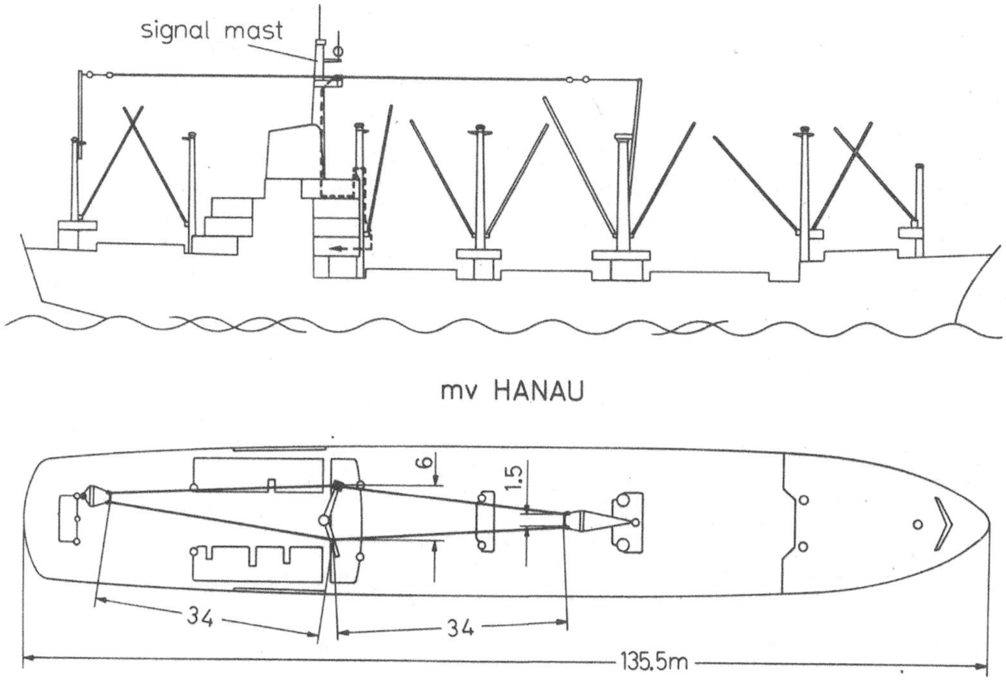

Fig. 14: Installation of the antenna on board of mv "Hanau"; the masts carrying the antenna are, on the left, a mast (8 m) especially added to the ship's mast, on the right, the heavy-lift derrick (60 tons), in the middle, the signal mast.

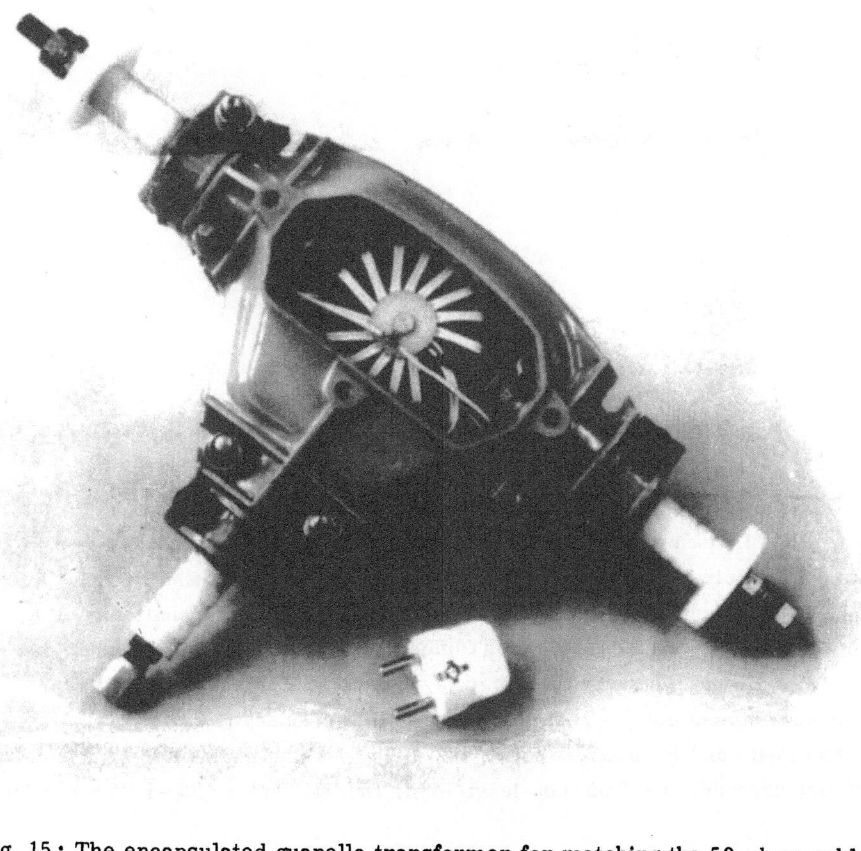

Fig. 15: The encapsulated guanella transformer for matching the 50-ohms cable to the symmetrical antenna feeding points. Bottom, left, the terminal for connection of the cable is shown, right, for comparison of sizes, a normal power-line plug.

Summary

In the period 28 November 1971 to 12 February 1972, during three passages of the North Atlantic, the diurnal variation of ionospheric absorption L was measured by method A1 at 1.7 and 5.0 MHz as a function of geographic and geomagnetic parameters. The measurements were carried out, applying a new equipment especially designed for mobile expeditions, on board a cargo vessel (mv "Hanau" of Hapag--Lloyd) moving between Hamburg (53° N; 10° E) and La Guaira (11° N; 67° W). Simultaneously, in the framework of the European Cooperation on Winter Anomaly studies, measurements of absorption were made at various other locations and the data used as a basis for comparison [circuits Norddeich-Lindau (A3; 2.61 MHz; 52.6° N; 8.7° E) and Aranjuez-Balerma (A3; 2.83 MHz; 38.6° N; 5.2° W), station Lancaster (A1; 2.14 MHz; 53°52' N; 2°45' W)]. Aboard, the diurnal variation of absorption showed large modifications from day to day. The behaviour of L ($\cos \chi = 0.2$) was somewhat different at each passage but showed a common trend: The winter anomaly vanished in a transit zone between 34° and 44° N centered at about 39° N; the same was true for the behaviour of L ($\cos \chi = 1$). Exponent n in $L = L_0 \cos^n \chi$ can be presented both as a function of geographic latitude and geomagnetic dip angle. An influence of the dip on the southern border of the winter-anomaly zone seems to be possible. The measurements made at 5 MHz, in general, supported those at 1.7 MHz. - A comparison of the trends of absorption at the three mentioned stations and the ship showed rather complicated and varying correlations.

Acknowledgements

The authors should like to acknowledge various kinds of support which were given them by a greater number of people of different institutions:

In the bureau of Hapag-Lloyd Captain Ulke, Mr. Lothar, Mr. Thomas, Mr. Hildebrandt, on the ship Captain Lüders, the 1st Officer Mr. Pietras, the 2nd Officers Mr. Arentz and Mr. Schmidt, the wireless operator Mr. Wanger, and the crew with Mr. Brusberg, who managed the up and down of the antenna, in the DEBEG (Deutsche Betriebsgesellschaft für drahtlose Telegraphie mbH, Member of the Radio Marine Associated Companies, London) Mr. Tobies, in the Fernmeldetechnisches Zentralamt (FTZ) der Deutschen Bundespost, Darmstadt, Mr. Richter, who sent and received necessary data and information by radiotelegram, a free of charge service by the FTZ, the Bundesministerium für das Post- und Fernmeldewesen, Bonn, for making available the frequencies and checking the degree of interference, and last not least Prof. Dr. W. Dieminger for his interest in this expedition. -

References

CAIN, J.C.	1963	J. Geophys. Res. 68, 4689
COYNE, T.N.R., BELROSE, J.S.	1972	Rad. Sci. 7, 163
DAVIES, K.	1965	National Bureau of Standards Monograph 80.
DIEMINGER, W., ROSE, G. WIDDEL, H.U.	1966	J. Atmos. Terr. Phys. 28, 317
FRITZ, S. SOULES, S.D.	1972	Mon. Wea. Rev. 100, 582
LAUTER, E.A. SCHÄNING, B.	1970	J. Atmos. Terr. Phys. 32, 1619
LABITZKE, K.	1972	J. Atmos. Sci. 29, 1395
LABITZKE, K. SCHWENTEK, H.	1968	Zs. Geophys. 34, 555
MAEDA, K.	1971	J. Geom. Geoel. 23, 133
SCHWENTEK, H.	1958	Arch. Elektr. Übertr. 12, 301
SCHWENTEK, H.	1966	Annls. Géophys. 22, 276
SCHWENTEK, H.	1968	Nachr. techn. Z. 21, 32
SCHWENTEK, H.	1971	J. Atmos. Terr. Phys. 33, 1647
WAKAI, N., OUCHI, Ch., NEMOTO, Ch.	1970	J. Rad. Res. Labs. Japan 17, 185
WRIGHT, J.W.	1963	J. Geophys. Res. 68, 4379

Reference is also made to the following material:

ELLING, W.	1968	Mitteilungen aus dem Max-Planck-Institut für Aeronomie, Nr. 36 (I), Springer-Verlag, Berlin, Heidelberg, New York
GRUSCHWITZ, E.H.	1972	Dissertation, Göttingen

Data used in this paper which were made available by other investigators

Ionospheric absorption at 2.14 MHz:	By courtesy of J.K. HARGREAVES, University of Lancaster, England
Ionospheric absorption at 2.83 MHz:	By courtesy of G. ROSE, J. WEBER and H.U. WIDDEL of the Max-Planck-Institut für Aeronomie, D 3411 Lindau, FRG

Verzeichnis der Mitteilungen aus dem Max-Planck-Institut für Physik der Stratosphäre

---

Nr. 1/1953 Über den Beitrag der von μ - Mesonen angestoßenen Elektronen zu den Ultrastrahlungsschauern unter Blei. G. Pfotzer

Nr. 2/1954 Ein Zählrohrkoinzidenzgerät zur Registrierung der kosmischen Ultrastrahlung. A. Ehmert

Eine einfache Methode zur Einstellung und Fixierung des Expansionsverhältnisses von Nebelkammern. G. Pfotzer

Nr. 3/1954 Optische Interferenzen an dünnen, bei -190^{0}C kondensierten Eisschichten. Erich Regener (vergriffen)

Nr. 4/1955 Über die Messung der Temperatur des atmosphärischen Ozons mit Hilfe der Huggins-Banden. H. Zschörner und H. K. Paetzold

Nr. 5/1956 Ein neuer Ausbruch solarer Ultrastrahlung am 23. Februar 1956. A. Ehmert und G. Pfotzer, vergriffen (erschienen Z. Naturforschung 11a, 322, 1956)

Nr. 6/1956 Das Abklingen der solaren Ultrastrahlung beim Ausbruch am 23. Februar 1956 und die geomagnetischen Einfallsbedingungen. A. Ehmert und G. Pfotzer

Nr. 7/1956 Die Impulsverteilung der solaren Ultrastrahlung in der Abklingphase des Strahlungseinbruches am 23. Februar 1956. G. Pfotzer

Nr. 8/1956 Die atmosphärischen Störungen und ihre Anwendung zur Untersuchung der unteren Ionosphäre. K. Revellio

Nr. 9/1956 Solare Ultrastrahlung als Sonde für das Magnetfeld der Erde in großer Entfernung. G. Pfotzer

*

Die vorstehenden Hefte können beim Max-Planck-Institut für Aeronomie, 3411 Lindau angefordert werden.

Mitteilungen aus dem Max-Planck-Institut für Aeronomie

Nr. 1 (S) 1959 Waibel: Messungen von Primärteilchen der kosmischen Strahlung.

Nr. 2 (S) 1959 Erbe: Auswirkung der Variationen der primären kosmischen Strahlung auf die Mesonen- und Nukleonenkomponente am Erdboden.

Nr. 3 (I) 1960 Kohl: Bewegung der F-Schicht der Ionosphäre bei erdmagnetischen Bai-Störungen.

Nr. 4 (I) 1960 Becker: Tables of ordinary and extraordinary refractive indices, group refractive indices and $h'_{o,x}(f)$-curves or standard ionospheric layer models.

Nr. 5 (S) 1961 Schröpl: Über eine Neubestimmung des Absorptionskoeffizienten von Ozon im Ultraviolett bei kleinen Konzentrationen.

Nr. 6 (S) 1961 Erbe: Ergebnisse der Ballonaufstiege zur Messung der kosmischen Strahlung in Weissenau und Lindau.

Nr. 7 (S) 1962 Meyer: Elektromagnetische Induktion eines vertikalen magnetischen Dipols über einem leitenden homogenen Halbraum.

Nr. 8 (I u. S) 1962 Dieminger und Mitarb.: Die geophysikalischen Ereignisse des 12. - 14. November 1960.

Nr. 9 (S) 1962 Pfotzer, Ehmert, and Keppler: Time Pattern of Ionizing Radiation in Balloon Altitudes in High Latitudes. Part A, Text; Part B, Figures and Diagrams.

Nr. 10 (S) 1963 Waibel: Eine Ballonsonde zur Messung von Röntgenstrahlung und solarer Ultrastrahlung.

Nr. 11 (S) 1963 Voelker: Zur Breitenabhängigkeit erdmagnetischer Pulsationen.

Nr. 12 (S) 1963 Jaeschke: Registrierung von Pulsationen im südlichen Niedersachsen als Beitrag zur erdmagnetischen Tiefensondierung.

Nr. 13 (S) 1963 Meyer: Elektromagnetische Induktion in einem leitenden homogenen Zylinder durch äußere magnetische und elektrische Wechselfelder.

Nr. 14 (S) 1964 Kremser: Über den Zusammenhang zwischen Röntgenstrahlungs-Ausbrüchen in der Polarlichtzone und bayartigen erdmagnetischen Störungen.

Nr. 15 (S) 1964 Keppler: Messung von Röntgenstrahlung und solaren Protonen mit Ballongeräten in der Nordlichtzone.

Nr. 16 (S) 1964 Kirsch: Die Anisotropien der kosmischen Strahlung.

Nr. 17 (S) 1964 Guilino: Ausbau eines Wechsellichtmonochromators und seine Anwendung zur Messung des Luftleuchtens während der Dämmerung und in der Nacht.

Nr. 18 (S) 1965 Pfotzer and Ehmert: Measurements of High Energetic Auroral Radiations with Balloon-Borne Detectors in 1962 and 1963 Part A to C, Text; Part D, Figures and Diagrams.

Nr. 19 (I) 1965 Hartmann: Bestimmung wichtiger Satellitenpositionen mit Hilfe graphischer Darstellungen.

Nr. 20 (S) 1965 Keppler: Über die Eigenschaften von Zählrohren und Ionisationskammern in verschiedenartigen Strahlungsfeldern. - Zur Interpretation von Röntgenstrahlungsmessungen in Ballonhöhe in der Nordlichtzone.

Nr. 21 (S) 1965 Siebert: Zur Theorie erdmagnetischer Pulsationen mit breitenabhängigen Perioden.

Nr. 22 (S) 1965 Meyer: Zur 27 täglichen Wiederholungsneigung der erdmagnetischen Aktivität, erschlossen aus den täglichen Charakterzahlen C 8 von 1884-1964.

Nr. 23 (S) 1965 Frisius: Über die Bestimmung von Längstwellen - Ausbreitungsparametern aus Feldstärkemessungen am Erdboden.

Nr. 24 (I) 1965 Ma: Einfluß der erdmagnetischen Unruhe auf den brauchbaren Frequenzbereich im Kurzwellen-Weitverkehr am Rande der Nordlichtzone.

Nr. 25 (S) 1965 Kremser, Keppler, Bewersdorff, Saeger, Ehmert, Pfotzer, Riedler, Legrand: X - Ray Measurements in the Auroral Zone from July to October 1964.

Nr. 26 (I) 1966 Stubbe: Theoretische Beschreibung des Verhaltens der nächtlichen F - Schicht.

Nr. 27 (S) 1966 Wilhelm: Registrierung und Analyse erdmagnetischer Pulsationen der Polarlichtzone, sowie ein Vergleich mit Bremsstrahlungsmessungen.

Nr. 28 (S) 1967 Fabian: Über eine neue Ozonradiosonde und Untersuchung von Lufttransporten in der unteren Stratosphäre.

Nr. 29 (S) 1967 Specht: Über die Absorptions- und Emissionsstrahlung der atmosphärischen Ozonschicht bei der Wellenlänge 9,6 μ.

Nr. 30 (I) 1967 Rose und Widdel: Ein Meßgerät zur Bestimmung der Strömungsgeschwindigkeit in kurzen Rohren (Ionenzählern) bei niedrigem Gasdruck.

Nr. 31 (I) 1967 Hartmann: Die Amplitudenregistrierungen des Satelliten Explorer 22, unter besonderer Berücksichtigung der Effekte, die bei Elevationswinkeln kleiner als 45° auftreten.

Nr. 32 (I) 1967 Rüster: Lösung von Bewegungsgleichungen und Kontinuitätsgleichung der F - Schicht mit speziellen Anwendungen auf erdmagnetische Baistörungen.

Nr. 33 (S) 1968 Müller: Zur Modulation der kosmischen Strahlung.

Nr. 34 (S) 1968 Münch: Statistische Frequenzanalyse von erdmagnetischen Pulsationen.

Nr. 35 (S) 1968 Schreiber: Das Magnetfeld des Ringstroms während der Hauptphase erdmagnetischer Stürme und ein Vergleich mit dem beobachteten D_{st}-Anteil des Störfeldes.

Nr. 36 (I) 1968 Elling: Spezielle Näherungsformeln der Appleton-Hartree-Gleichungen zur Interpretation der Absorption einer Mittelwellenausbreitung im nächtlichen E-Gebiet der Ionosphäre.

Nr. 37 (I) 1968 Jones: Application of the Geometrical Theory of Diffraction to Terrestrial L F Radio Wave Propagation.

Nr. 38 (S) 1969 Zürn: Zum weltweiten Auftreten erdmagnetischer Pulsationen vom Typ pc 4.

Nr. 39 (S) 1969 Tiefenau: Untersuchungen an Kanal-Elektronen-Vervielfachern.

Nr. 40 (S) 1970: Sonderheft zum 60. Geburtstag von Herrn Prof. Dr.-Ing. G. Pfotzer am 29. November 1969 und Herrn Prof. Dr.-Ing. A. Ehmert am 6. März 1970.

Nr. 41 (S) 1970 Stratmann: Berechnung des Wellenfeldes eines Längstwellensenders im Entfernungsbereich bis 1000 km zur kontinuierlichen Sondierung der tiefen Ionosphäre durch Feldstärkemessungen in geeigneten Entfernungen vom Sender.

Nr. 42 (S) 1970 Pruchniewicz: Über ein Ozon-Registriergerät und Untersuchung der zeitlichen und räumlichen Variationen des Troposphärischen Ozons auf der Nordhalbkugel der Erde.

Nr. 43 (S) 1970 Richter: Über eine Ballonsonde für Polarlichtmessungen und über den Vergleich von Polarlichtemissionen, Röntgenstrahlen und ionosphärischen Absorptionen.

Nr. 44 (S) 1970 Niapour: Untersuchungen über die mittlere Multiplizität der Verdampfungsneutronen als Maß für die Veränderungen des Energiespektrums der kosmischen Strahlung.

Nr. 45 (S) 1971 Tiefenau: Messungen von Ozonprofilen über dem Meer und Bestimmung des Ozonflusses in die Meeresoberfläche sowie der spezifischen Ozonzerstörungsrate in der maritimen Grenzschicht.

Nr. 46 (S) 1972 Roeckner: Temperaturberechnung der Venusatmosphäre bis 80 km Höhe aufgrund solarer und thermischer Strahlungsströme sowie konvektiver und turbulenter Wärmetransporte.

Nr. 47 (S) 1972 Holl: Zur Theorie thermisch angeregter Gezeiten in der E-Schicht der Ionosphäre.

Nr. 48 (I) 1972 Hartmann, Oberländer, Schmidt, Schödel: Satellite Beacon Observations from 1964 to 1970.

Nr. 49 (S) 1972 Stüdemann: Direkte Teilchenmessungen im Morgensektor der Polarlichtzone.

Nr. 50 (S) 1973 Jessen: Ein Rechenmodell zur Beschreibung des stratosphärischen Ozonkreislaufs.

GPSR Compliance

The European Union's (EU) General Product Safety Regulation (GPSR) is a set of rules that requires consumer products to be safe and our obligations to ensure this.

If you have any concerns about our products, you can contact us on ProductSafety@springernature.com

In case Publisher is established outside the EU, the EU authorized representative is:

Springer Nature Customer Service Center GmbH
Europaplatz 3
69115 Heidelberg, Germany

Batch number: 10684220

Printed by Printforce, the Netherlands